There's a chance that you know how to swim. But Walter and I didn't. We couldn't go in deep water because it wasn't safe. I'll tell you what we're doing about it now. We're taking swimming lessons!

Ms. Small is our teacher. She's a person who's able to help kids swim. Even if you're shy or don't like water, she'll get you swimming! She's always saying, "You'll love to swim once you've learned how."

Ms. Small also tells us, "You won't learn all about swimming in just one day. You shouldn't expect to. This isn't a race, so what's the rush? We'll all learn to swim one step at a time."

There are many parts to swimming, and the first one's breathing. Walter knows what to do. He'll take a breath while his head's out of water. Then he'll blow out air when his head's in water. See how he's doing it?

I wasn't able to float at first, but now I haven't any problem! It doesn't take much to relax my body, so I don't sink or fall. Wouldn't you just love to float on water all day long?

Kicking is a key part of swimming. You've got to keep your legs going so they'll move you in the water. They've got to move up and down quickly. My legs weren't that strong at first, but now they're very strong!

You've also got to use both arms in swimming. Each one's like a paddle. Walter's arms always move in the water. He'll pull one arm forward while he's pushing one arm back. That's what I call swimming!

Ms. Small says we're swimming well. But that doesn't mean we shouldn't keep getting better. She's right. I'll tell you what she's also right about. You'll love to swim once you've learned how!